Maximizing ChatGPT

Unleashing the full potential of ChatGPT in E-commerce

Mark O'Niel

Table of Contents

An Introduction to GPT-3 and ChatGPT

In the world of natural language processing, one model that stands out for its capability and performance is ChatGPT. Developed by OpenAI, ChatGPT is a large language model that can generate human-like text, understand natural language, and complete a wide range of language tasks.

ChatGPT (short for "Conversational Generative Pre-training Transformer") is a large language model developed by OpenAI. It is based on the GPT-3 architecture, which has been trained on a massive amount of text data, making it one of the most powerful language models available. The model is designed to generate human-like text, understand natural language, and complete a wide range of language tasks.

One of the most significant advantages of ChatGPT is its ability to generate human-like text. This feature makes it an ideal tool for creating chatbots, product descriptions, and product recommendations. With ChatGPT, you can train a chatbot to respond to customer inquiries in a natural and human-like manner. Also, you can use ChatGPT to generate product descriptions that are engaging, informative, and SEO-friendly. In addition, you can train ChatGPT to recommend products to customers based on their browsing history, purchase history, and other factors.

Another advantage of ChatGPT is its ability to understand natural language. This feature makes it an ideal tool for tasks such as sentiment analysis, text classification, and text summarization. With ChatGPT, you can analyze customer feedback, social media posts, and reviews to understand customer sentiment. Also, you can use ChatGPT to classify text into different categories, such as product reviews, customer service inquiries, and so on. In addition, you can train ChatGPT to summarize text, such as customer reviews, to quickly understand key takeaways.

By the end of this book, you will have a clear understanding of how to use ChatGPT for natural language processing tasks, and you will be able to implement the model for various use cases in e-commerce. ChatGPT is a powerful and versatile model for natural language processing (NLP) developed by OpenAI. It is based on the GPT (Generative Pre-trained Transformer) architecture and is trained on a massive amount of text data to generate human-like text. Some of the key capabilities of ChatGPT for natural language processing include:

1. **Language generation**: ChatGPT can generate text in a variety of languages and styles, such as creative writing, poetry, song

lyrics, script writing, speeches, etc. This is achieved by training the model on a large dataset of text in the target language and fine-tuning it for specific tasks.

2. Language Translation: ChatGPT can translate text from one language to another by training it on a large dataset of parallel text in the source and target languages, and fine-tuning it for specific tasks. The model can understand the meaning of the input text and generate a corresponding translation in the target language.

3. Text summarization: ChatGPT can generate a shorter version of a text that contains the most important information. This is achieved by training the model on a large dataset of text and fine-tuning it for specific tasks. This capability can be useful in areas such as news, legal and medical documents, academic papers, etc.

4. Text completion: ChatGPT can generate text that continues a given prompt, by training on a large dataset of text and fine-tuning it for specific tasks. This capability can be useful in areas such as coding, songwriting, screenplays, email writing, etc.

5. Question answering: ChatGPT can answer questions by understanding the meaning of the input text and generating a corresponding answer. This is achieved by training the model on a large dataset of text and fine-tuning it for specific tasks.

6. Dialogue systems: ChatGPT can generate human-like responses in a dialogue by understanding the context of the conversation and generating a corresponding response. This is achieved by training the model on a large dataset of dialogues and fine-tuning it for specific tasks.

7. **Sentiment analysis**: ChatGPT can classify text as positive, neutral, or negative by understanding the meaning of the input text and generating a corresponding label. This is achieved by training the model on a large dataset of text and fine-tuning it for specific tasks.

8. **Text-to-Speech and Speech-to-Text**: ChatGPT can be used to convert text to speech, and speech-to-text text by fine-tuning it for specific tasks and providing it with the right dataset. This capability can be useful in areas such as voice assistants, call centers, etc.

9. **Text data augmentation**: ChatGPT can be used to generate new text based on a given prompt, which can be useful in areas such as data cleaning, data visualization, data exploration, data analysis, data summarization, data reporting, etc.

10. Language understanding: ChatGPT can understand the meaning of the input text by training on a large dataset of text and fine-tuning it for specific tasks, this capability can be useful in areas such as machine reading comprehension, natural language inference, etc.

These are just some examples of the many capabilities of ChatGPT for natural language processing. The model's ability to generate human-like text, understand the meaning of a text, and generate corresponding output makes it a powerful tool for a wide range of NLP tasks. However, it's worth noting that the model is not perfect and still has some limitations such as bias, lack of common sense, and the need for a large amount of data and computational resources to

Fine-tuning ChatGPT for specific tasks

Collect a labeled dataset: The first step in fine-tuning ChatGPT is to collect a labeled dataset that is specific to the task or domain you want to fine-tune the model for. The dataset should contain input-output pairs that are relevant to the task or domain, and it should be large enough to train the model effectively.

Preprocess the dataset: The next step is to preprocess the dataset by cleaning and tokenizing it. You can use various preprocessing techniques such as lowercasing, removing stopwords, and so on. It is also important to make sure that the input-output pairs match in format and that the dataset is shuffled.

Fine-tune the model: Once you have a preprocessed dataset, you can fine-tune the model by training it on the dataset. You can use various fine-tuning techniques such as discriminative fine-tuning, adversarial fine-tuning, and more. It is also important to select an appropriate batch size and learning rate and to monitor the training progress.

Evaluate the model's performance: Once the model is fine-tuned, you can evaluate its performance by inputting test examples from the task or domain and comparing the model's output to the ground truth. You can use various metrics such as accuracy, BLEU score, and so on.

Fine-tune further: If the model's performance is not satisfactory, you can fine-tune the model further by adjusting the hyperparameters, providing more data, or using a different fine-tuning technique.

Use the model: Once you are satisfied with the model's performance, you can use it for the specific task or domain.

Please keep in mind that fine-tuning a model like ChatGPT requires a lot of computational resources and a good amount of labeled data. Also, it is important to note that fine-tuning a model like ChatGPT does not change the model architecture but rather adjusts the weights of the model to better fit the new data. It is also important to monitor for bias and ensure that the fine-tuned model respects data privacy.

Specific Use Cases of ChatGPT

Here are a few specific use cases of ChatGPT in e-commerce:

Chatbot for Customer Service

ChatGPT can be used to train a chatbot that can respond to customer inquiries in a natural and human-like manner. The model can be fine-tuned to create chatbots that can understand and respond to customer inquiries in a natural and human-like manner. The model can be fine-tuned for tasks such as intent recognition and response generation.

Intent recognition is the process of identifying the intent behind a customer's inquiry. For example, a customer's inquiry "Can you help me track my order?" has the intent of tracking an order. Intent recognition is a crucial task for chatbots as it allows the chatbot to understand the customer's inquiry and respond appropriately.

Response generation is the process of generating a response for a customer's inquiry. Once the chatbot has identified the intent behind the customer's inquiry, it can generate a response that is appropriate for that intent. For example, if the customer's inquiry has the intent of tracking an order, the chatbot can generate a response that provides the customer with the status of their order.

To fine-tune ChatGPT for chatbot tasks such as intent recognition and response generation, you will need to provide the model with a labeled dataset that contains customer inquiries and corresponding intents and responses. This dataset can be used to fine-tune the model so that it can accurately identify intents and generate appropriate responses.

Once the model is fine-tuned, it can be integrated into a chatbot platform such as Dialogflow or Microsoft Bot Framework. The chatbot can then be deployed on different channels such as a website, mobile app, or messaging platform.

It's important to note that creating chatbots with ChatGPT requires a good amount of labeled data and computational resources. Also, it is important to handle common challenges such as data privacy and bias, by removing any sensitive information from the data and also ensuring that the dataset used for fine-tuning the model is diverse and representative of the population that the chatbot will be serving.

Best practices for creating chatbots with ChatGPT include:

1. Collecting a diverse and representative dataset that covers a wide range of customer inquiries and intents.
2. Regularly updating the dataset and fine-tuning the model to adapt to changes in customer inquiries and intents.
3. Implementing a way to handle out-of-scope inquiries, such as redirecting the customer to a human agent or providing a list of frequently asked questions.
4. Monitoring the performance of the chatbot and using metrics such as accuracy and customer satisfaction to identify areas for improvement.
5. Regularly testing the chatbot with a diverse group of users to identify any issues with bias or fairness.

Real-world examples of chatbots created with ChatGPT

1. A chatbot for a retail company that can assist customers with tasks such as tracking orders, finding products, and making returns.
2. A chatbot for a bank that can assist customers with tasks such as checking account balances, making payments, and reporting lost or stolen cards.
3. A chatbot for a customer service department that can assist customers with tasks such as troubleshooting technical issues and answering frequently asked questions.

ChatGPT can be fine-tuned to create chatbots that can understand and respond to customer inquiries in a natural and human-like manner. By providing the model with a labeled dataset that contains customer inquiries and corresponding intents and responses, and implementing best practices such as regularly monitoring the performance and testing with a diverse group of users, organizations can create chatbots that are accurate, efficient, and fair.

Product Recommendations

ChatGPT can be used to train a model that can recommend products to customers based on their browsing history, purchase history, and other factors. The model can be fine-tuned to recommend products to customers based on their browsing history, purchase history, and other factors. The model can be fine-tuned for tasks such as product similarity and user-product matching.

Product similarity is the process of identifying products that are similar to a given product. For example, a customer is viewing a product "red dress" and the model can recommend similar products such as "red dress with different patterns" or "red dress with different lengths".

User-product matching is the process of recommending products to a customer based on their browsing history, purchase history, and other factors. For example, a customer who frequently buys red dresses may be recommended products such as "red high heels" or "red purse".

To fine-tune ChatGPT for product recommendation tasks, you will need to provide the model with a dataset that contains information about products and customers. This dataset can be used to fine-tune the model so that it can accurately identify similar products and recommend products to customers based on their browsing history and purchase history.

Once the model is fine-tuned, it can be integrated into a recommendation system that can be used to recommend products to customers on an e-commerce website or mobile app.

It's important to note that creating product recommendations with ChatGPT requires a good amount of data and computational resources. Also, it is important to handle the cold-start problem by providing a default set of recommendations for new customers or customers with little browsing history.

Best practices for creating product recommendations with ChatGPT

1. Collecting and analyzing a large dataset that contains information about products and customers, including browsing history and purchase history.
2 Regularly updating the dataset and fine-tuning the model to adapt to changes in customer preferences and the product catalog.
3. Using metrics such as click-through rate (CTR) and conversion rate to evaluate the performance of the recommendations and identify areas for improvement.

4. Experimenting with different recommendation algorithms such as collaborative filtering and content-based filtering to find the best approach for your specific use case.
5. Incorporating additional information such as customer demographics and product attributes to improve the accuracy of the recommendations.

Real-world examples of product recommendations created with ChatGPT

1. A recommendation system for a fashion e-commerce website that recommends clothing and accessories to customers based on their browsing history and purchase history.
2. A recommendation system for a streaming service that recommends movies and TV shows to customers based on their viewing history and ratings.

3. A recommendation system for a food delivery service that recommends menu items to customers based on their order history and dietary restrictions.

ChatGPT can be fine-tuned to recommend products to customers based on their browsing history, purchase history, and other factors. By providing the model with a dataset that contains information about products and customers, and implementing best practices such as regularly monitoring performance and experimenting with different recommendation algorithms, organizations can create recommendations that are accurate, personalized, and effective.

Product Descriptions

ChatGPT can be used to generate product descriptions that are engaging, informative, and SEO-friendly. ChatGPT can be fine-tuned to generate product descriptions that are engaging, informative, and SEO-friendly. The model can be fine-tuned for tasks such as product attribute extraction and text generation.

Product attribute extraction is the process of identifying important product attributes such as color, size, and material from a product's title, image or text. These attributes can be used to generate more informative and accurate product descriptions.

Text generation is the process of generating a product description using the extracted product attributes. Once the model has identified the important product attributes, it can generate a product description that is engaging, informative, and SEO-friendly.

To fine-tune ChatGPT for product description tasks such as product attribute extraction and text generation, you will need to provide the model with a labeled dataset that contains product attributes and corresponding product descriptions. This dataset can be used to fine-tune the model so that it can accurately extract product attributes and generate appropriate product descriptions.

Once the model is fine-tuned, it can be integrated into an e-commerce platform or CMS to automatically generate product descriptions for new products.

It's important to note that creating product descriptions with ChatGPT requires a good amount of labeled data and computational resources. Also, it is important to handle common challenges such as data privacy and bias, by removing any sensitive information from the data and also ensuring that the dataset used for fine-tuning the model is diverse and representative of the population that the product descriptions will be serving.

Best practices for creating product descriptions with ChatGPT

1. Collecting a diverse and representative dataset that covers a wide range of products and corresponding product attributes.
2. Regularly updating the dataset and fine-tuning the model to adapt to changes in product attributes and trends in product descriptions.
3. Monitoring the performance of the generated product descriptions, using metrics such as readability, informativeness and search engine optimization.
4. Regularly testing the generated product descriptions with a diverse group of users to identify any issues with bias or fairness.
5. Incorporating customer feedback and reviews when fine-tuning the model to ensure the generated product descriptions accurately reflect the customer's perspective.

Real-world examples of product descriptions created with ChatGPT

1. An e-commerce platform that uses ChatGPT to generate product descriptions for a wide range of products, improving the informativeness and readability of the descriptions while also increasing the platform's SEO.

2. A fashion retailer that uses ChatGPT to generate product descriptions for their clothing line, highlighting the key features and attributes of each product in an engaging and informative manner.

3. A home goods retailer that uses ChatGPT to generate product descriptions for their furniture line, providing customers with detailed information on the materials, dimensions, and styles of each product.

ChatGPT can be fine-tuned to generate product descriptions that are engaging, informative, and SEO-friendly. By providing the model with a labeled dataset that contains product attributes and corresponding product descriptions, and implementing best practices such as regularly monitoring the performance and testing with a diverse group of users, organizations can create product descriptions that are accurate, informative, and engaging.

Sentiment Analysis

ChatGPT can be used to analyze customer feedback, social media posts, and reviews to understand customer sentiment. The model can be fine-tuned on a dataset of customer feedback and reviews, and it can learn to understand the tone and sentiment of the text, and classify it as positive, negative, or neutral. This can help e-commerce businesses understand customer sentiment about their products and services, and make improvements accordingly. Sentiment

analysis can be useful for understanding customer sentiment and identifying areas for improvement in products or services.
To fine-tune ChatGPT for sentiment analysis, you will need to provide the model with a labeled dataset that contains text and corresponding sentiment labels (positive, negative, neutral). This dataset can be used to fine-tune the model so that it can accurately predict the sentiment of a given text.

Once the model is fine-tuned, it can be integrated into a sentiment analysis platform or used as a standalone tool to analyze customer feedback, social media posts, and reviews.

It's important to note that fine-tuning ChatGPT for sentiment analysis requires a good amount of labeled data and computational resources. Also, it is important to handle common challenges such as data privacy and bias, by removing any sensitive information from the data and also ensuring that the dataset used for fine-tuning the model is diverse and representative of the population that the sentiment analysis will be serving.

Best practices for sentiment analysis with ChatGPT

1. Collecting a diverse and representative dataset that covers a wide range of sentiments and topics.
2. Regularly updating the dataset and fine-tuning the model to adapt to changes in customer sentiment and trends in the language used to express it.
3. Monitoring the performance of the sentiment analysis, using metrics such as accuracy and F1-score.
4. Regularly testing the sentiment analysis with a diverse group of users to identify any issues with bias or fairness.
5. Incorporating customer feedback and reviews when fine-tuning the model to

ensure the sentiment analysis accurately reflects the customer's perspective.

Real-world examples of sentiment analysis created with ChatGPT

1. A customer service department that uses ChatGPT to analyze customer feedback and identify areas for improvement in their products or services.
2. A social media monitoring tool that uses ChatGPT to analyze posts and comments on social media platforms and identify sentiment towards a brand or campaign.
3. A review analysis platform that uses ChatGPT to analyze customer reviews on e-commerce websites and identify the overall sentiment towards a product or service.

ChatGPT can be fine-tuned for sentiment analysis, which is the process of determining the attitude or emotion of a piece of text, such as customer feedback, social media posts, and reviews. By providing the model with a labeled dataset that contains text and corresponding sentiment labels, and implementing best practices such as regularly monitoring the performance and testing with a diverse group of users, organizations can create sentiment analysis models that are accurate and unbiased.

In addition, by regularly updating the dataset and fine-tuning the model, organizations can ensure that the sentiment analysis accurately reflects changes in customer sentiment and trends in language. Sentiment analysis can be useful for understanding customer sentiment and identifying areas for improvement in products or services.

CONCLUSION

In this book, we have discussed the capabilities of ChatGPT for natural language processing and its potential use in e-commerce. Specifically, we have looked at how ChatGPT can be used to create chatbots, generate product recommendations, generate product descriptions, perform sentiment analysis.

We've provided an overview of how ChatGPT can be used for each of these tasks, including the technical setup and requirements, a step-by-step guide on how to fine-tune ChatGPT, best practices, and real-world examples and case studies.

In terms of creating chatbots, we've seen that ChatGPT can be fine-tuned to understand customer inquiries and respond in a natural and human-like manner. This can be useful for providing customers with accurate and efficient assistance in areas such as order tracking and customer support.

In terms of product recommendations, we've seen that ChatGPT can be fine-tuned to recommend products to customers based on their browsing history, purchase history, and other factors. This can be useful for increasing sales and customer satisfaction by providing personalized product recommendations.

In terms of product descriptions, we've seen that ChatGPT can be fine-tuned to generate product descriptions that are engaging, informative, and SEO-friendly. This can be useful for improving the online shopping experience and increasing the visibility of products in search engines.

In terms of sentiment analysis, we've seen that ChatGPT can be fine-tuned to determine the attitude or emotion of a piece of text, such as customer feedback, social media posts, and reviews. This can be useful for understanding customer sentiment and identifying areas for improvement in products or services.

In terms of future directions, there are many other potential use cases for ChatGPT in e-commerce, such as:

Using ChatGPT to generate product labels and tags for improved searchability and product discovery.
Using ChatGPT to generate product comparison charts and tables for improved product comparison and decision making.
Using ChatGPT to generate product tutorials and instructions for improved product understanding and usage.
Using ChatGPT to generate product FAQs and knowledge bases for improved customer support.
We have seen that ChatGPT is a powerful tool for natural language processing and has a wide range of potential applications in

e-commerce. With the right approach and best practices, organizations can use ChatGPT to improve customer experience, increase sales, and drive business growth.
A walk-through of the most common use cases for ChatGPT such as language translation, text summarization, text generation, dialogue systems and more.

www.ingramcontent.com/pod-product-compliance
Lightning Source LLC
LaVergne TN
LVHW011822240225
804421LV00004B/631

* 9 7 9 8 3 7 4 7 0 6 7 2 7 *